DATE DUE		
AUG 11 2000		
DEC 27 2000		
FEB 16 2001	JAN 2 9 2014 NVP.	
MAR 0 6 2001		
	JUL 0 6 2015	
MAR 25 2002		
APR 2 9 2002		
AUG 2 8 2002	DEC 0 5 2015	
OCT 1 0 2002	JAN 0 4 2016	
JAN 17 2004	MAR 1 8 2016	
MAR 1 6 2008	APR 2 0 2016	
AUG 12 2013		
APR 2 3 2014		
APR 3 0 2015		

Anne S. Bittker
Children's Collection

NEW HAVEN
FREE PUBLIC
LIBRARY

BETCHA!

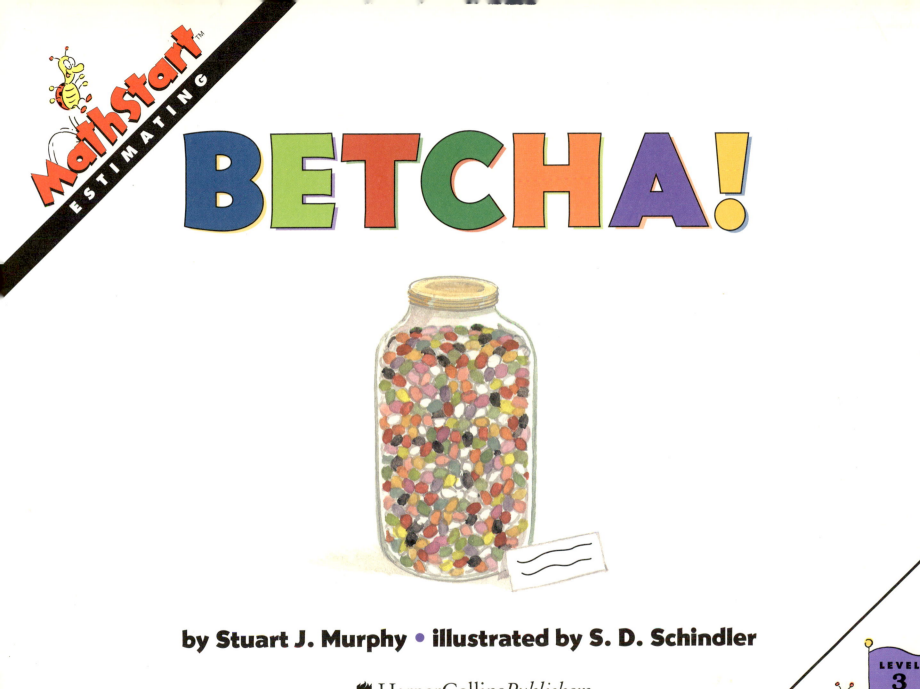

by Stuart J. Murphy • illustrated by S. D. Schindler

HarperCollins*Publishers*

**To Katherine Brown Tegen—
who was willing to betcha
that MathStart would work.**

—S.J.M.

For more information about the MathStart series, please write to
HarperCollins Children's Books, 10 East 53rd Street, New York, NY 10022,
or visit our web site at http://www.harperchildrens.com.
Bugs incorporated in the MathStart series design were painted by Jon Buller.
HarperCollins®, ■®, and MathStart™ are trademarks of HarperCollins Publishers Inc.

BETCHA!

Library of Congress Cataloging-in-Publication Data
Murphy, Stuart J., date.
  Betcha! / by Stuart J. Murphy ; illustrated by S. D. Schindler.
     p.   cm. — (MathStart)
  "Level 3, Estimating."
  Summary: Uses a dialog between friends, one who estimates, one who counts precisely, to show
estimation at work in everyday life.
  ISBN 0-06-026768-2. — ISBN 0-06-026769-0 (lib. bdg.). — ISBN 0-06-446707-4 (pbk.)
  1. Estimation theory—Juvenile literature.   [1. Estimation theory.   2. Arithmetic.]   I. Schindler,
S. D., ill.   II. Title.   III. Series.
QA276.8.M87   1997   96-15486
519.5′44—dc20   CIP
                                                                                       AC

Typography by Elynn Cohen
1 2 3 4 5 6 7 8 9 10
❖
First Edition

BETCHA!

Hey! Look at this. Planet Toys is having a contest. Whoever guesses the correct number of jelly beans in the jar in their window wins two free tickets to the All-Star Game!

PLANET TOYS

Guess how many jelly beans are in me!

Oh, yeah?
Let's go down and check it out.

Betcha I can win.

Oh yeah? I'm really good at figuring things out.
I betcha that I'll win.

If you're so good at figuring things out, can you tell me how many other people are on this bus?

You betcha!

9

Okay. How many?

4 people ⟶ X

10 rows

= 40, plus a few standing up

Betcha about forty-three people.

1, 2, 3 . . . 42, 43, 44, 45!

I just counted, and I got forty-five.

I was pretty close!

Wow! Look at that traffic jam.
Betcha you can't tell me how many cars are
stuck on the block.

Betcha I can.

Go ahead and try.

6 cars

4 lanes

X

= 24

Betcha about twenty-five cars.

17

1, 2, 3 . . . 21, 22, 23!

I counted twenty-three.

I was almost right!

Here's the store. Look at all that cool stuff! About how much do you think it would cost to buy it all?

PLANET TOYS

$39

WALKIE-TALKIES

$28

Turbo-Fire

$22

$12

Betcha I can tell you about how much.

Go for it!

$39

WALKIE-TALKIES

$22

Turbo-Fire

$28

Almost $40 + just over $20 = about $60

$60 + almost $30 = about $90

$90 + just over $10
= about $100

$12

Betcha about $100.

23

I get $101.

See? I was close again,
and I didn't even need a pencil.

Okay. Now for the real thing.
About how many jelly beans are in the jar?

Guess how many jelly beans are in me and win two free tickets to the All-Star Game?

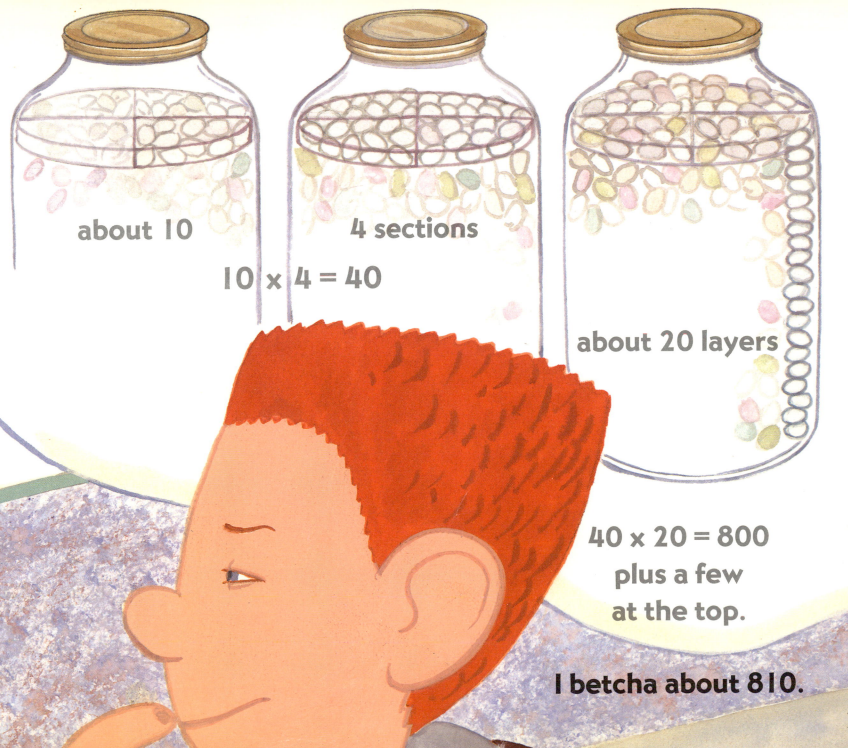

about 10

4 sections

$10 \times 4 = 40$

about 20 layers

$40 \times 20 = 800$
plus a few
at the top.

I betcha about 810.

Well, I can't count them—and
he's been pretty close every time.
I'll just add a few to his number.

Incredible! That's exactly right!
You're headed to the All-Star Game.

Betcha I know who wants the other ticket!

If you would like to have more fun with the math concepts presented in *Betcha!*, here are a few suggestions:

- Read the story together and ask the child to describe what is going on in each picture. Ask questions throughout the story, such as "What would you do to estimate how many people are on a bus?" and "How would you estimate the number of cars in a traffic jam?"

- Help the child to understand what it means to estimate: Ask "How did you get your answer?" and "Is it about right?" Encourage the child to reconsider the estimate as more information is available. The process is more important than an exact answer.

- Discuss real-life situations that require estimations. Examples could include ordering enough pizza for the whole family or deciding how many errands can be done before ballet class or soccer practice.

- Try using estimation to check against calculator work. Estimate the total cost of four or five items in the refrigerator. Then add them up on a calculator. Was the estimate a reasonable answer? Was it about right? Try again with another group of items.

- Together, make up your own "Betcha!" game. Pick something that is difficult to count, such as people in a long line, cars in a parking lot, or cookies in a box. Help the child to consider different strategies for making these estimates. Then check to see how close these estimates are to the real numbers.

Following are some activities that will help you extend the concepts presented in *Betcha!* into a child's everyday life.

Eating Out: Estimate the number of people eating at a favorite restaurant. Do a quick count to see how close the estimate is. Try to estimate how many tables there are. How many chairs? About how much will the check total?

Taking a Walk: Estimate as you walk around your neighborhood. If you pass so many houses on one block, about how many houses will you pass after walking all the way to the park? If it takes a certain number of steps to walk between two lines on the sidewalk, about how many steps will it take to walk the entire length of the street? Around the block?

Buying Food: Estimate how much your family spends each month on one popular food item—cereal, for example. About how many boxes does your family eat each month? About how much does each box cost? About how much do you spend? Then keep track and see if your estimate is close.

The following books include some of the same concepts that are presented in *Betcha!*:

- Counting on Frank by Rod Clement

- Counting Jennie by Helena Clare Pittman

- How Much Is a Million? by David M. Schwartz

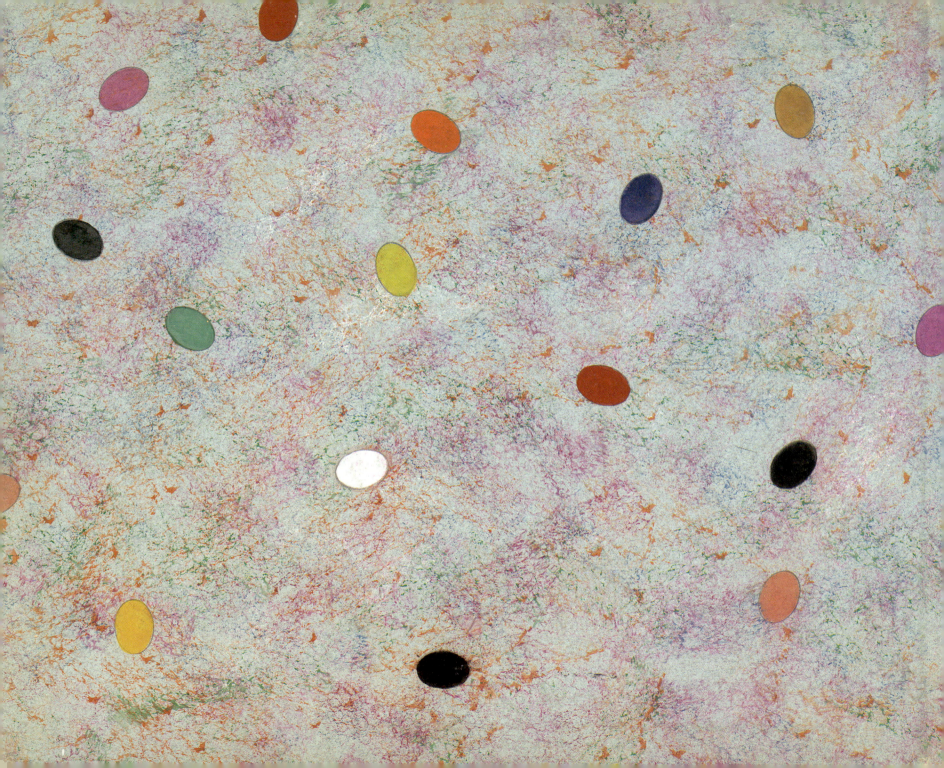